above the rest

EUROFIGHTER TYPHOON

A PICTORIAL HISTORY BY GEOFFREY LEE

Published in 2007 by Ad Hoc Publications
Cedars, Wattisham Road, Ringshall, Suffolk IP14 2HX, England
www.adhocpublications.com

Copyright © Geoffrey Lee, 2007

The right of Geoffrey Lee to be identified as the author of this book is asserted in accordance with the Copyright, Patents and Designs Act 1988.

All rights reserved. No part of this publication may be reproduced or transmitted in any form whatsoever without the prior written permission of the Publisher.

ISBN 978 0 946958 69 6

Designed by Andreas Westphal, images.art.design.
www.iad-design.de

Printed and bound in Germany by ESTA DRUCK GMBH
www.esta-druck.com

Contents

5	**Preface**
6	**Pilots' Views** Eurofighter Typhoon test pilots
8	**Introduction** From concept, design and first flights to full production
18	**Production** Manching, EADS Deutschland, Germany Caselle, Alenia Aeronautica, Italy Getafe, EADS-CASA, Spain Warton, BAE Systems, United Kingdom
26	**Development Aircraft** DA 1 to DA 7 IPA 1 to IPA 6
66	**German Air Force** Fighter Wings JG 73 and JG 74, based at Laage/Rostock and Neuburg/Donau
102	**Italian Air Force** IX°, 12° and XX° Squadrons, based at Grosseto and Gioia del Colle
120	**Spanish Air Force** Wing (Ala) 11, based at Morón de la Frontera
136	**Royal Air Force** 3 (F), XI, 17 (R) and 29 (R) Squadrons, based at RAF Coningsby
176	**Austrian Air Force** Surveillance Wing, based at Zeltweg/Steiermark
192	**Acknowledgements**

Preface

I have always had a fascination with aircraft and a passion for photography, and I have been very fortunate over the last 28 years to have combined both in my career.

I started my career in the aerospace industry with Hawker Siddeley, based in Kingston-upon-Thames in Surrey. Hawker Siddeley was then one of the largest aviation businesses in the United Kingdom (it has now been absorbed into BAE Systems). I joined as trainee photographer – and I could not believe my luck!

I can honestly say that I have never looked back. The day I realised that I was in my element was the day I carried out my first air-to-air sortie. I remember vividly the excitement and adrenaline rush as, airborne, I photographed a Hawk on a local test flight from Dunsfold airfield. I discovered that I had an aptitude for air-to-air photography, and I thrived on the challenge of capturing images of aircraft being put through their paces.

My first encounter with Eurofighter Typhoon, as a functioning aircraft, was in April 1994 when I photographed the first British-built development aircraft, DA 2, during its first flight from Warton Aerodrome. As the programme developed, Eurofighter Typhoon became the subject of many of my photographs.

Because of this, I have had the privilege and honour of working with the leading test pilots and air force pilots of the four partner nations in the programme, Germany, Italy, Spain and the United Kingdom. These pilots continue to inspire me to capture creative images that demonstrate the agility and capability of this fantastic aircraft.

Eurofighter Typhoon is probably the most versatile combat aircraft currently in service anywhere in the world, and I count it a tremendous honour to be associated with it. **Above the Rest** is a tribute to the aircraft and to the many teams who have contributed to its design and development.

The images in the pages that follow capture the enormity of the journey Eurofighter Typhoon has made thus far, from its birth, through its development and to its service with the air forces of the four partner nations and that of Austria, and the challenge in compiling the book lay in trying to do justice to that journey. I hope that you will enjoy the story as much I have enjoyed the experience of photographing this magnificent machine.

I would like to dedicate this book to my wife, Melanie, for being so supportive during my assignments around Europe in pursuit of that experience.

Geoffrey Lee

Pilots' Views

I have been testing Eurofighter Typhoon actively for over ten years. The aircraft combines all the attributes that are really important to a fighter pilot. The combined power and lift of the Eurofighter Typhoon gives it an equal performance to that of the American super-fighter, the F-22, but with a much smaller size and at a fraction of the cost. At the same time, utilising external fuel tanks, Eurofighter Typhoon can match the range capability of potentially the most advanced US fighter-bomber under development, the F-35, while maintaining an advantage in performance and weapons load.

Apart from firepower and capability, safety is critically important, particularly when on active deployment. In combat, the pilot needs advanced self-protection systems to avoid being shot down. Eurofighter Typhoon has these in abundance, and, in conjunction with the twin-engine design and dual redundancy on practically all aircraft systems, they make it one of the safest and most reliable jets ever.

The need for extra inches to house the large, powerful antenna has brought about a welcome increase in dimensions to the cockpit, making it one of the largest and most comfortable fighter cockpits built. It has also been my privilege to display the aircraft at numerous international air shows. It is a very complex weapon system, but the pilot interface makes it extremely easy to operate. It is not surprising, then, that the aircraft is such a joy to fly. It has been my most rewarding professional experience to be part of the development of the Eurofighter Typhoon.

Chris Worning
EADS-Deutschland
Experimental Test Pilot
Eurofighter Typhoon Project Pilot

On a cool autumn morning in October 1996, I took off from Caselle Airport, Italy, for my first flight at the controls of Eurofighter Typhoon Development Aircraft 3 (DA 3) as OTC Test Pilot. It may have been eleven years ago, but it feels like yesterday. For me, it was the beginning of a beautiful adventure and of the very rewarding experience of being deeply involved in test flying the Eurofighter Typhoon.

This weapon system is a great flying machine – the dream of every pilot. Impressive performance, superb flying qualities and powerful sensors all combine to give the exciting feeling of being "on top of the world". For me, it was love at first sight, and I have never been let down. I still get surprised and pleased by the aircraft with every flight; there is not much more that a test pilot can ask for. Every time I step out to go and fly, I realise that I am very lucky to be part of the Eurofighter Typhoon team, sharing in such a wonderful adventure.

Marco Venanzetti
Alenia Aeronautica
Chief Test Pilot

Without a doubt, the Eurofighter Typhoon is the aircraft of my life. It was love at first sight, from the moment I first touched it, sitting down in the cockpit and greedily exploring the interior of this marvellous machine. The aircraft is the offspring of human imagination and effort. However, when I pushed forward the throttles and released the brakes, and when my helmet flattened against the seat and my heart began to beat faster, I realised that I had to make some effort if I wanted to reach the level of my new partner.

Some first flights that I have made in the past have been full of misunderstandings, but with Eurofighter Typhoon all my expectations were exceeded. As my time working with the aircraft continued, all the most advanced systems, offering me precisely what I needed and when I needed it, were slowly introduced into it. On the other hand, the performance of the aircraft has taken me to the outer edges of my physical limits. We have both learned so much from each other.

Today, our relationship is full of revelations and, with every flight, Eurofighter Typhoon demonstrates an improvement – and yet I know that there are so many things still to be discovered. I really fear not being able to keep up with the challenge and becoming more of an obstacle to an aircraft that is now capable of demonstrating to the entire aeronautics world that its time has come. The pilot and Eurofighter Typhoon form a unique partnership.

Alfonso de Castro
EADS-CASA
Chief Test Pilot Eurofighter Typhoon

It was the summer of 2002 at Warton Aerodrome in Lancashire, England, when I first descended the boarding ladder of single-seat Eurofighter Typhoon DA 2; I had just completed my first test sortie in this remarkable aircraft. The crew chief looked first over the aircraft and then over at me and said, "Looks like you've got the 'Eurofighter Typhoon grin', then!" The Eurofighter Typhoon grin was something I would get used to. Whether it was the customer pilots, test pilot colleagues or even dignitaries, their broad smiles from ear to ear meant that we had a real world-beater on our hands.

What is it that makes this aircraft so special for me? Could it be the advanced cockpit, stuffed with the latest technology? Is it the revolutionary Voice Throttle and Stick (VTAS) design? This is a feature whereby the pilot can not only operate the weapon system and its multitude of smart sensors without removing his hands from the controls, but also "talk" to the aircraft. Maybe it is the Helmet Mounted Display (HMD), which, in combination with an advanced short-range missile such as ASRAAM, remains perfectly suited for any close-in dogfight, irrespective of the adversary being in front, to the side or even over the shoulder. It could also be the phenomenal agility and handling qualities of the aircraft. A "carefree" digital flight control system lets the pilot do almost anything with the stick, whilst the flight control computers wring every last ounce of aerodynamic and inertial performance out of the airframe. The truth is that it is all of these and more.

The same control laws that provide the awesome agility are also in place to protect the pilot, whether the aircraft is in the clean configuration or fully loaded with weapons. In my aerial display at the Farnborough International 2006 air show, where the aircraft carried six 1,000lb (450kg) Paveway laser-guided bombs, a mixture of six air-to-air missiles and a centreline fuel tank, throwing the aircraft about remained completely natural, safe and, quite simply, stunning. Performance like this, of course, needs a good combination of airframe design and engine. The EJ200 engine not only provides the performance margins to the Eurofighter Typhoon, but also allows the pilot carefree control of the throttles down to the lowest of airspeeds.

Both engine and airframe are still very much in their youth, and there is much more to come from this European collaborative project. I am proud of what we have achieved so far, but I also know that, with continued effort, technical know-how and structured development, Eurofighter Typhoon will remain one of aerospace's success stories. From a test pilot's perspective it is quite simply – awesome!

Mark Bowman
BAE Systems
Chief Test Pilot Eurofighter Typhoon

Eurofighter Typhoon Background

During the 1970s, when new American combat aircraft such as the F-15 and the F-16 – and, more importantly, the new Soviet fighter designs, the MiG-29 and the Su-27 – appeared, several European air forces were confronted with the fact that their own fighter fleets were beginning to lag behind. As a consequence, they had to work quickly in order to define a suitable answer.

The Panavia Tornado programme, a collaboration involving Germany, Italy and the United Kingdom, was already in progress at the time, but, although the various prototypes had flown and conversion units had been established, the aircraft was not due to become operational before the end of the decade. Meanwhile, existing types such as the Harrier, Jaguar, F-4 Phantom II and F-104 continued in service. Even as the Tornado programme was under development, however, the Royal Air Force identified a need for a future aircraft programme to succeed it, and, accordingly, an Air Staff Target (AST) was issued.

This AST called for a short take-off, vertical-landing aircraft to supplant the Germany-based Harriers and Jaguars. By the middle of the 1970s the requirements had altered, and a new AST was issued to include a secondary air superiority role for the design. Having already established a partnership with France (SEPECAT) and with Germany and Italy (Panavia), the United Kingdom approached France and Germany with a view to sharing the development costs of the proposed new aircraft. However, the requirements of the three nations differed considerably, and discussions were in progress for a number of years. Eventually, in 1979 an agreement was reached to begin a two-year feasibility study for a European Combat Fighter (ECF).

European Designs

British Aerospace (BAe) submitted their P.106 and P.110 proposals, Messerschmitt-Bölkow-Blohm (MBB) offered their TKF-90 and Dassault produced their ACX, but, since the designs differed fundamentally, an agreement could not be reached and the proposals were shelved. With the success of Tornado assured, the three Panavia partners then initiated an Agile Combat Aircraft (ACA) programme, but this, too, was destined to proceed no further than a full-scale replica (which was shown at the Farnborough International Air Show in 1982 and at Le Bourget in Paris the following year). Interestingly, the mock-up aircraft had many features that would eventually be seen on the Eurofighter Typhoon.

By now time was running out and the requirement had become urgent. It was at this stage that the United Kingdom Government stepped in and funded an Experimental Aircraft Programme (EAP), signing an £80 million contract in May 1983. Further funding was supplied by BAe, MBB and Aeritalia (Italy) in order to build a single demonstrator aircraft. Design work was shared among the partnership and the construction of the aircraft was commissioned to BAe's facility at Warton. Meanwhile a new Future/European Fighter Aircraft (F/EFA) programme had been launched with France and Spain joining the established partnership, but agreement on design and build sharing could not be reached, France demanding leadership of the entire programme. As a result, France left the partnership in July 1985 in order to pursue her own combat aircraft design (which resulted in Rafale).

On 16 May 1983 the United Kingdom Ministry of Defence awarded BAe and Aeritalia a contract for one EAP, with the expectation that Germany would quickly commit to the construction of a second demonstrator (MBB having intimated that it would provide the rear fuselage elements of the EAP). However, when the funding was cut, BAe simply used the rear section of a Tornado, including a modified version of the tailfin. This rear section was made mostly of aircraft alloys, but the rest of the airframe consisted largely of graphite-epoxy composite assemblies. The aircraft incorporated a quadruple-redundant fly-by-wire Flight Control System (FCS), which was a necessity as the EAP demonstrator was intentionally designed to be aerodynamically unstable – meaning that it would quickly depart from the controlled flight envelope unless the onboard computers continuously performed tiny adjustments to the flight control surfaces. Although EAP required many lines of software, this aerodynamic instability helped give the aircraft high agility. Power was supplied by twin Turbo-Union RB199 afterburning turbofans as fitted in the Tornado. As with the Agile Combat Aircraft, the air intakes were placed under the belly, and these had a hinged panel on the lower lip that could be dropped open to ensure that a satisfactory airflow was delivered to the engines at high angles of attack.

Experimental Aircraft Programme

The EAP continued independently of the F/EFA and, in October 1985, the prototype (ZF534) was unveiled to the world's press. Many design elements of the ACA had been adapted and had found their way on to the aircraft, which made its first flight on 8 August 1986.

The EAP demonstrator featured a "glass cockpit", with three Smiths Industries multifunction displays using colour cathode ray tubes, a GEC-Marconi wide-angle Head-Up Display (HUD) and centre-mounted Hands-on-Throttle-and-Stick (HOTAS) controls. BAe incorporated a voice-warning system in the cockpit and also studied a Direct Voice Input (DVI) command system for the aircraft. Many of the suggestions for the cockpit layout and fit had come from the test pilots who had been an essential part of the design team, and, as a consequence, the cockpit was widely regarded as outstanding and extremely pilot-friendly. After some 259 test flights, EAP was retired on 1 May 1991 and donated to the Department of Aeronautical Studies at Loughborough University, where it continues to inspire a new generation of engineers and designers.

With the withdrawal of France from the F/EFA project, Germany, Italy, Spain and the United Kingdom were left in the frame, but their various requirements were still some way apart: Spain and the United Kingdom were looking for a multi-role fighter, whereas Germany and Italy were seeking an air superiority fighter with no air-to-ground capability. However, although EFA was focused on air superiority, it could perform ground attack as a secondary mission. As the design evolved, EFA was to have high performance, high manoeuvrability and carefree handling characteristics. It would also have a low radar cross-section (RCS) and be capable of operating from short forward airstrips.

Discussions commenced and an agreement was reached in June 1986 when Eurofighter GmbH, an industrial consortium made up of the established aerospace companies Aeritalia (now Alenia Aeronautica) in Italy, British Aerospace (now BAE Systems) in the UK, Construcciones Aeronáuticas SA (now EADS-CASA) in Spain and MBB (now EADS) in Germany was formed. At the same time, the four partner Governments had agreed to set up their own body, the NATO Eurofighter and Tornado Development, Production and Logistics Management Agency (NETMA), to manage the programme on their behalf and represent the single interface to the industrial consortium.

Having already built Tornado as a partnership, Germany, Italy and the United Kingdom had an established and proven framework in place with which to work. Building the aircraft had to take into account the various industrial complexities, and a suitable workshare programme for the development phase was next on the agenda. This shared construction work was allocated in proportion to the planned orders – Germany and the United Kingdom with 33 per cent each, Italy with 21 per cent and Spain with 13 per cent. The four countries issued a final requirement for the proposed fighter, and a development contract for the airframe and engine was signed in November 1988. Alongside the brand new aircraft, a new engine, the EJ200, was designed and built by the Eurojet partnership, which comprised MTU (now MTU Aero Engines) of Germany, Fiat Avio (now Avio SpA) of Italy, SENER (now ITP) of Spain and Rolls-Royce of the United Kingdom.

The arrangements concerning the design and production of Eurofighter Typhoon are relatively straightforward. The "single source" principle has been applied, and the employment of integrated teams assures full access to, and exchange of, technology among the partners. As the production workshare is allocated according to the number of aircraft ordered, the taxpayers' money spent by the nations stays "at home" and thus contributes to the overall economy and social system of each country. This system also avoids cross-invoicing among the nations. Eurofighter GmbH is contracted to NETMA and the consortium places orders with the four industrial partner companies, thereby allowing the aircraft's configuration to be managed throughout production and facilitating the introduction of the latest engineering developments into the programme without compromising either manufacture or in-service support.

Each of the partner companies has System and Design Responsibility for different aspects of the programme and can therefore issue sub-contracts to other suppliers.

British Aerospace's ACA mock-up at Warton

British Aerospace's Experimental Aircraft Programme (EAP) technology demonstrator was built to test the design features of the proposed Eurofighter. The aircraft, ZF534, made its first flight on 8 August 1986.

Workshare and System Design Responsibility

Alenia Aeronautica, Italy:
- Utilities Control System
- Wing Design
- Navigation Sub-System
- Engine Integration
- Secondary Power System
- Fuel System
- Role Equipment (50 per cent)

BAE Systems, United Kingdom:
- Avionics Integration
- Front and Rear Fuselage
- ECM, Lightning Tests
- Electrical System
- Fuel System
- Defensive Aids Sub-System
- Role Equipment (50 per cent)

EADS-CASA, Spain:
- Structure Technology
- Wing Design
- Main Airframe Test
- Environmental Control System
- Communications Sub-System

EADS, Germany:
- Flight Control System
- Attack and Identification Sub-System
- Centre Fuselage Design
- Main Airframe Test
- Radar Signature
- Hydraulic Systems
- Landing Gear System
- Gun Integration

The allocation of design and production work highlights the international flavour of the entire programme. Besides the main partners, there are over 400 other companies across the supplier network involved in the Eurofighter project. Once the workshare had been decided, the actual task of constructing the aircraft could begin, and work on the prototypes commenced in late 1989. The first Eurofighter Typhoon was completed in May 1992.

Eurofighter Typhoon Début

What could not have been foreseen was how the drama of major world events at the end of the 1980s would change the political landscape. Since the initial discussions and commitments to the Eurofighter Typhoon project, the Berlin Wall had been torn down and what had been known as the Iron Curtain had fallen and the Cold War had come to an end. The Communist threat had evaporated almost overnight. This meant that the original role specified for the Eurofighter Typhoon was in doubt. Costs had spiralled, and decisions taken by Germany, faced with additional expenditure stemming from reunification, almost caused the entire programme to come to a halt. The Germans demanded a trimmed-down version of Eurofighter Typhoon; otherwise, they would pursue the option of purchasing aircraft "off the shelf" from the United States. They raised a number of serious objections, proposing, for example, that the new fighters carry an improved version of the AN/APG-65 radar in service on the US F/A-18 Hornet instead of a European system. They also suggested that early development aircraft use General Electric F404 engines instead of Turbo-Union RB199s. That made no sense to the other partners, as the European RB199 was in service and easily available.

DASA: DA 1's first flight, Manching, 27 March 1994

Discussions between the partners followed, and a reorientation phase lasting almost two years was undertaken, during which a thorough assessment of the new requirements and the original specification was made. Eventually, differences were resolved and Germany opted to stay with the Eurofighter Typhoon, advising that it would cut its order from 250 to 140 aircraft (this was later increased to 180). The other partners also pruned their original orders, though to a lesser degree. Production workshares were finally recalculated: the United Kingdom was allocated 37 per cent (232 aircraft), Germany 30 per cent (180), Italy 20 per cent (121) and Spain 13 per cent (87).

This brought the programme back on track, but Germany's procrastination had meant that the planned in-service date for 1995 would not now be met. Slow funding from the German Government, its delay in signing the production investment contract for several months and its general lack of commitment seriously delayed and indeed jeopardised the programme. When the investment contract was finally signed, the programme was eleven months behind the production schedule and the in-service date had slipped. It must be said that during this time, even though the politicians had been vociferous about not requiring Eurofighter Typhoon, the German Air Force had all along insisted that they wanted the aircraft. This united voice certainly helped to persuade the decision-makers.

British Aerospace: DA 2's first flight, Warton, 4 April 1994

Once the partner nations had sorted out their differences, and with funding firmly in place, the programme took on a new lease of life and went ahead with renewed effort – and with a new name, EF2000, to signify the revised date for the aircraft's entry into service. Later, in September 1998, the name Typhoon was chosen for EF2000s destined for service in the Royal Air Force and for export outside Europe; elsewhere, the use of this name was left to the discretion of the member nations. The title EF2000 was subsequently abandoned and never again referred to by industry, although the Governments continued to use it for some time afterwards.

Alenia Aeronautica: DA 3's first flight, Caselle, 4 June 1995

British Aerospace: DA 4's first flight, Warton, 14 March 1997

DASA: DA 5's first flight, Manching, 24 February 1997

CASA: DA 6's first flight, Getafe, 31 August 1996

First Flights

Nine prototype or development aircraft were ordered, but two, one each from Spain and the United Kingdom, were later cancelled. Using the data gathered by EAP, work continued on the Development Aircraft (DA) in each of the partner countries. The Development Aircraft were assigned specific flight test responsibilities and the resulting information was contributed to the four-nation flight test programme.

The first Development Aircraft to fly, designated DA 1 (and coded 98+29), was built by DASA and flown from Manching by test pilot Peter Weger on 27 March 1994. It did not carry any radar and was powered by twin Turbo-Union RB199 engines as used by Tornado. After the initial general proving flights and some checks on the flight control software, DA 1 was subjected to ground running trials until September that year. After a short pause, during which time it was fitted with upgraded software for the Flight Control System, DA 1 resumed flying.

Across in the United Kingdom, on 4 April 1994 BAe chief test pilot Chris Yeo took control of DA 2 (ZH588) for its maiden flight from Warton. Like DA 1, this aircraft was also fitted with RB199 engines and, also like the first aircraft, was grounded following initial flight trials and flight control systems tests. By this time, the EJ200 flight engines were ready: the first pair were fitted to the Italian DA 3, which made its début on 4 June 1995, piloted by Napoleone Bragagnolo from Turin-Caselle. DA 3 was used for engine development trials and underwing fuel tank tests; like its companions, this Eurofighter Typhoon was not fitted with radar. The next aircraft to fly was the twin-seat DA 6 (XCE.16-01), built in Spain and flown by Alfonso de Miguel Gonzalez on 31 August 1996. Joining its three colleagues on the test programme, DA 6 was used for envelope expansion, for climatic and Environmental Control System (ECS) testing and for Multifunction Information Distribution System (MIDS) and helmet development. Unfortunately, DA 6 was lost on 21 November 2002 following a double-engine flameout.

Just under five months after DA 6 first flew, the test programme was joined on 27 January 1997 by DA 7 (MMX603), built by Alenia Aeronautica; once again, it was Napoleone Bragagnolo who made the first flight. DA 7 was used for navigation, communications, flight performance, weapons integration and firing trials. It was the first Eurofighter Typhoon to fly armed with the AIM-9L Sidewinder missile and the AIM-120 Advanced Medium Range Air-to-Air Missile (AMRAAM), and it was later was fitted with Passive Infra-Red Airborne Track Equipment (PIRATE). Although it was the final aircraft in the programme, DA 7 was not the

Alenia Aeronautica: DA 7's first flight, Caselle, 27 January 1997

last to join the team. Next came the German-built DA 5 (98+30), which made its first flight on 24 February 1997 piloted by Wolfgang Schirdewahn. This aircraft was the first to be fitted with Euroradar ECR90 (now known as Captor). It was used for radar and avionics trials and was also the first of the development fleet to fly outside the airspace of the partner nations when it conducted demonstration flights in Norway in 1998.

The last Eurofighter Typhoon Development Aircraft to enter the programme was BAe's two-seat DA 4 (ZH590), which made its first flight on 14 March 1997 under the control of test pilot Derek Reeh after a delay brought about by the need to install the Captor radar and a complete avionics suite. DA 4 soon entered its phase in the test programme, carrying out weapons integration and sensor fusion trials as well as radar tests. Eurofighter Typhoon DA 4 made the first guided AMRAAM firing and, on 20 February 1998, was also the first aircraft in the programme to "supercruise", achieving over Mach 1.0 on dry power (i.e., without the use of reheat). When the units became available, the earlier aircraft were re-engined with EJ200s in place of the RB199s and the test programme continued apace as the Development Aircraft fulfilled their Main Development Contract objectives.

Three-Tranche Production Programme

At the end of 1997, the four partner nations, having finally secured funding, signed an inter-governmental Memorandum of Understanding covering Production Investment and Production Logistics Support. One month later, on 30 January 1998, the Supplement 1 Production Investment, Production and Support contracts were signed covering tooling and preparatory work for the production of 620 aircraft plus 90 options. This "Umbrella Contract" for Eurofighter Typhoon divided production into three tranches, with capabilities progressively enhanced as new systems and weapons became available, ensuring that the aircraft remained at the forefront of modern technology. Delivery numbers for the three tranches were agreed as 148 for Tranche 1 and 236 for both Tranche 2 and Tranche 3.

Tranche 1 was divided into two batches, each corresponding to one or two different levels of weapon system specification. Batch 1 was subdivided into Block 1, Block 1B and Block 1C. These blocks differed considerably one from the other as the main design was modified according to data from the development programme and in response to the customer's requirements. Batch 2 was subdivided into Blocks 2, 2B, and 5, each needing a different software suite and different electronic line-replaceable units. Unlike the Batch 1 modifications, these variations did not alter the physical appearance of the aircraft. With two planned upgrade programmes, all the aircraft of Tranche 1 are to be brought to the Final Operational Capability as specified in the Main Development Contract. The first upgrade, Retrofit R1, has already been concluded, bringing the Batch 1 aircraft to the hardware standard of Batch 2; the second upgrade programme, Retrofit R2, which will bring all early aircraft up to the Block 5 capability, is well under way, the first such example having been delivered in 2007.

Of the 148 aircraft making up Tranche 1, 53 are for the United Kingdom (35 single-seaters and 18 two-seaters, including a fatigue test airframe), 42 will go to Germany (26 and 16), 28 will be delivered to Italy (18 and 10) and Spain will receive 19 (11 and 8). These figures were adjusted slightly in July 2007 following the renegotiation of the procurement contract with the Republic of Austria, whose air force will receive Tranche 1 Block 5 standard aircraft, nine from EADS Military Air Systems' final assembly line at Manching and six from the R2 upgrade programme. These aircraft were first intended for the German Air Force, who instead will receive the Tranche 2 aircraft that were originally to be handed over to Austria.

Because of the advanced nature of Eurofighter Typhoon, the complex and exhaustive test-flying programme was crucial to the success of the aircraft. Following the Development Aircraft, the next test assets were the Instrumented Production Aircraft (IPA). Of these there are now six, with a seventh scheduled to have joined the test fleet by the end of 2007. The first to fly was IPA 2 (MMX614), from Italy, on 5 April 2002 – although IPA 1 and IPA 3 made their first flights within ten days of IPA 2 becoming airborne. These three aircraft were thrust straight into flight testing in order to provide the clearance and certification for Type Acceptance of the Batch 1 standard Eurofighter Typhoon.

The tests for each particular application were not limited to a single chosen aircraft: many of the trials were shared among the three IPAs, with data from the different sources compared.

The 30 Batch 1 aircraft were all twin-seaters and were delivered with the Initial Operational Capability (IOC) standard avionics software, tailored for crew training, air defence training and role evaluation. All were armed with the Mauser BK27 27mm cannon, could carry four AIM-120B AMRAAMs, two AIM-9Ls or ASRAAMs and external fuel tanks and were equipped with the Captor radar, and thus had the basic air-to-air and air defence capabilities, but they lacked Defensive Aids Sub-Systems. It is worth mentioning here that every Eurofighter Typhoon is fitted with the BK27 cannon, mounted in the starboard wing root. The weapon was originally to have been a feature of all aircraft except those destined for the Royal Air Force, but the RAF reassessed it and decided to make use of it. DA 3 was the first to fire the cannon, on 13 March 2002, and exercises with service aircraft have included gun firing on various weapons ranges.

Among the production aircraft in the programme, two were singled out to become Instrumented Series Production Aircraft (ISPA), one each for Italy and the United Kingdom. Although these were owned and operated by the customer nations, they were used to confirm and augment data gathered by the DA and IPA test fleets.

Type Acceptance

On 30 June 2003 the Eurofighter Typhoon achieved international Type Acceptance for the first aircraft. This signature involving NETMA and Eurofighter GmbH represented the final stage in clearance to allow the delivery of Batch 1 twin-seat aircraft to the four partner air forces, and the handover of the first aircraft paved the way for pilot instructor training and conversion to type to begin.

Industry, meanwhile, were striving to introduce the Batch 2 single-seat variant into service. The next Instrumented Production Aircraft, IPA 4 and IPA 5, fitted with full IOC standard avionics software and enhanced air defence capabilities, had made their first flights by mid-2004. Armament options were expanded to allow the full use of AIM-9L Sidewinder, ASRAAM, AIM-120B AMRAAM and IRIS-T. They were also fitted with Captor radar (which included a friend-or-foe identification system), and with a basic Defensive Aids Sub-System that embraced electronic support measures, some countermeasures and chaff/flare dispensers. Also included were experimental microwave landing systems, Initial Direct Voice Input and a limited autopilot system. Eurofighter Typhoon had come a long way since the days of EFA.

Type Acceptance was achieved on 13 December 2004, and the next day, 14 December, the Tranche 2 Supplement 3 contract was signed by NETMA and Eurofighter GmbH for the production of a further 236 aircraft, giving the consortium the largest order book for any new fighter aircraft worldwide. Deliveries of the first single-seat Batch 2 aircraft began to get under way, and by April 2005 all four air forces were operating them. Deliveries of the final Tranche 1 standard, Block 5 aircraft began in March 2007 with the handover of SS011 to the Spanish Air Force. This latest standard opens up the multi-role potential of Eurofighter Typhoon, offering an air-to-ground capability to complement that of air supremacy. The "Austere Capability" contract, signed in July 2006, represents the go-ahead for the integration of a Laser Designator Pod (LDP) into Royal Air Force Block 5 aircraft, thereby offering a genuine multi-role capability to that service's front-line squadrons. Negotiations for the signature of the Tranche 3 production contract will take place during 2008.

Displaying four-nation markings, Eurofighter Typhoon IPA 3 98+03 takes off from Manching

Export Success

In July 2002 the Eurofighter consortium were granted exclusive contract negotiations with the Austrian Government to meet the aircraft requirements of the Austrian Air Force, and in August 2003 a contract for 18 Eurofighter Typhoons was finalised and Austria joined the programme as the first export customer. However, political developments in that country forced a renegotiation. This was concluded in June 2007, the new agreement covering the delivery of 15 aircraft with the latest capability standard of Tranche 1. The handover of the first aircraft took place in July 2007.

A second success was achieved when the Governments of Saudi Arabia and the United Kingdom signed a contract for the supply of 72 Eurofighter Typhoon aircraft to the Saudi Arabia Kingdom. Announcement of the contract in September 2007 inaugurated Project Salam, ushering in a new chapter in the long and successful programme of co-operation in the defence field between the two countries.

Type Acceptance ceremony in Manching: (from left to right) Dr Peter Struck of Germany; General Poyo Guerrero of NETMA; On. Filippo Berselli of Italy; Excmo. St. Victor Torre de Silva of Spain; Filippo Bagnato, CEO of Eurofighter; and Lord William Bach of Lutterworth from the UK.

EJ200 – The Power behind Eurofighter Typhoon

An element that gave the Eurofighter consortium an edge was the fact that they were starting with a clean sheet. As already noted, there were trials and tribulations to be overcome, but the basic partnership framework was in place. With Tornado, there was in existence a very good engine, the RB199, which had been the work of a Rolls-Royce-led group. This unit was acceptable for the application it had been designed for, but the new fighter required an all-new propulsion system to push the boundaries of engine performance.

In 1982 Rolls-Royce commenced work on a new engine for the United Kingdom Ministry of Defence. Given the code XG-40 Advanced Core Military Engine, or ACME demonstrator, the programme was split into three phases – technology (1982–88), engine (1984–89) and assessment (1989–95). For XG-40, Rolls-Royce developed a new fan, compressor, combustor, turbine (including high temperature life prediction) and augmentor systems using advanced materials and new manufacturing processes. The first full engine commenced rig testing in December 1986, the final XG-40 running for some 200 hours during 4,000 cycles to bring the programme to a close in June 1995.

While the XG-40 was being designed and built, the German, Italian, Spanish and United Kingdom partners were busy with the design of the Eurofighter Typhoon airframe. They required a suitable engine – one that would complement their design. Encouraging tests of the XG-40 ably demonstrated that it was the engine for their aeroplane. Consequently, a new company, Eurojet Turbo GmbH, was formed to develop and build the new engine. Suitably, Eurojet consists of a consortium of companies from each of the partner nations. The shareholder percentages are: Rolls-Royce of the United Kingdom, 33 per cent; MTU Aero Engines of Germany, 33 per cent; Fiat Avio of Italy, 21 per cent; and ITP (Industria de Turbo Propulsores) of Spain, 13 per cent.

On the formation of the Eurojet consortium in 1986, much of the continuing XG-40 research was used for the new programme. Eurofighter Typhoon required a powerplant that was capable of higher thrust and longer life, and was less complex, than previous engines. This resulted in an engine with similar dimensions to the proven Rolls-Royce RB199, yet having almost half as many parts (1,800 against 2,845 for the RB199) and delivering nearly 50 per cent more dry thrust. The new engine was given the code EJ200. A major area in which the EJ200 differs from the older engine is in turbine blade technology.

The EJ200 is an advanced design based on a fully modular augmented twin-spool low bypass layout, while the compressor utilises a three-stage Low Pressure Fan (LPF) and a five-stage High Pressure Compressor (HPC). The EJ200 fan features wide-chord bladed/disc (blisk) assemblies designed for low-weight, high-efficiency operation.

By using a very low bypass ratio (the ratio of air that bypasses the compressor stages), the EJ200 has a near-turbojet cycle. This low ratio gives the engine the added benefit of producing a cycle where the maximum attainable non-afterburning thrust makes up a greater percentage of total achievable output. It is interesting to note that the baseline production engine is also capable of generating a further 15 per cent dry thrust and 5 per cent reheat. However, excessive use of this capability will result in a reduced life expectancy. The engine has been designed for a life of 6,000 hours, corresponding to about 30 years of operational use.

The technology of the EJ200 engine makes it both smaller and simpler in layout than current powerplants of a similar thrust class, while giving it lower fuel consumption and an unprecedented power-to-weight ratio – vital factors that contribute greatly to the multi-mission performance and effectiveness of the Eurofighter Typhoon. The advanced aerodynamics employed in the fan mean that optimum operation is achieved without the need for inlet guide vanes. All specified ingestion tests (for example, the ingestion of birds, hail and ice) were carried through without any problems affecting the structure and performance of the engine.

Throughout the engine, brush seals are widely used instead of labyrinth seals in the air system. The annular combustor, which incorporates air spray fuel injectors, has been designed for extremely low smoke and emission characteristics. The reheat system features radial hot stream burners and independent cold stream burning, and the engine boasts a hydraulically operated convergent/divergent nozzle.

These days, much is made of "supercruise" ability. Unlike some of the US manufacturers, the Eurofighter consortium has never greatly publicised this aspect of the aircraft's performance. However, during early flight test programmes Eurofighter Typhoon was found to be easily capable of supercruise, and this was written into the manuals. Service pilots are able to make good use of it and, in quoted figures, the Eurofighter Typhoon is capable of cruising at Mach 1.2 at 11,000m without reheat for extended periods. EADS has stated that a maximum upper limit of Mach1.5 is possible using the same engine, but that it would require a changed nozzle design. The ability to maintain transonic and supersonic flight regimes, without resorting to the use of reheat, has been made possible by the design and use of advanced materials in the engine. Eurofighter Typhoon can employ reheat with an upper (design) limit of Mach 2.0, when required. EJ200 has a thrust range from 13,500lbf dry maximum to 20,000lbf maximum with reheat.

To facilitate easier maintenance, Eurojet has incorporated many labour-saving devices. The Engine Health Monitoring System (EHMS) comprehensively monitors the engine's health and allows "on condition"' maintenance, without the need periodically to remove and service the engine. In-built Engine Health Monitoring design features remove the need for fixed intervals for off-wing engine overhauls. On Tranche 2 engine variants, the Engine Monitoring Unit (EMU) has been combined with the architecture of the digital engine control unit, removing the requirement to have an independent monitoring unit housed within the airframe. The combined unit is known as the Digital Engine Control and Monitoring Unit (DECMU).

The EJ200 engine provides comprehensive access for boroscope inspections of the complete gas path, including the combustor. The On-board Oil Debris Monitoring System complements Magnetic Chip Detectors, which can be used in conjunction with a commercially available debris analysis system. In order to make the maximum use of life-limited parts, the Engine Monitoring System calculates cyclic life consumption related to mission profile and engine usage. All accessories, including the digital engine control unit, are self-contained and engine-mounted. An auxiliary gearbox on the underside of the engine provides drive for the accessories. Because the engine has been designed to give equal priority to performance and life-cycle cost, emphasis was given to producing a reliable and easily maintainable modular unit with 15 fully interchangeable modules.

Eurojet is currently exploring possible enhancements and believes that there is a good opportunity for continuing investment in engine technology. The performance of a military aircraft engine not only is a significant factor in overall operational effectiveness but also affects the availability and life-cycle costs of the weapon system, and continuous improvement must be a consideration in enhancing that weapon system. Of course, it is not just about thrust, which is instantly recognisable to any pilot, but also about fuel burn (influencing reach, persistence and operational costs), all aspects of supportability, and component life (availability and cost of ownership).

With the EJ200 engine, a superior product design concept at low operational cost has been achieved by using Europe's centres of excellence and know-how. To assure the competitiveness of the EJ200 today and to secure its position in future markets, enhancements need to be considered. Eurojet's co-operative venture delivers significant benefits to both customer and participating industry. Industrial benefits are obtainable as a result of shared costs and the combined strengths of technological power and know-how.

The distinctive jet pipe of the EJ200 engine. Twenty-four petals make up the nozzle and have been nicknamed "turkey feathers".

Each Eurofighter aircraft is fitted with two Eurojet EJ200 engines, developed in partnership in Europe.

Production

Final assembly facility of the Eurofighter Typhoon at Manching for the German and Austrian Air Forces

EADS Deutschland – Manching, Germany

EADS in Germany is responsible for 33 per cent of all Eurofighter Typhoon production and the final assembly of the 180 Eurofighter Typhoons destined for the German Air Force (work that began in December 2000) and the 15 for the Austrian Air Force.

The final assembly facility at Manching is located 40km north of Munich. The airfield was opened in 1939 as a Luftwaffe flying school and has been a military air base throughout its life. The German Air Force Flight Test Centre, WTD 61, is located at Manching and a small commercial terminal was opened in the 1970s.

EADS in Germany has overall design responsibility for the Eurofighter Typhoon centre fuselage, the production of which involves several EADS plants in northern Germany that manufacture rear structural components and the carbon-fibre skin. These include the Military Air Systems factories in Augsburg, where the centre and forward parts of the fuselage section are manufactured and assembled.

The first Eurofighter Typhoon, DA 1, made its first flight at Manching on 27 March 1994 and the first series production aircraft for Germany took off on its maiden flight in February 2003.

Manching also serves as an alternative base for fighter and fighter-bomber units in southern Germany, for example during runway repairs at their home bases, and as a base for special international ventures such as the annual "ELITE" Electronic Warfare exercise.

Alenia Aeronautica – Caselle, Italy

Alenia Aeronautica is responsible for 21 per cent of all Eurofighter Typhoon production. This includes the final assembly of the 121 aircraft for the Italian Air Force at their production and flight test facilities at Turin-Caselle Airport, Italy.

The history of the site dates back to the 1930s, when increased military activities called for the construction of a new and larger airport, far from the city centre. Caselle Airport, which lies north of Turin between the foothills of the Alps and the Po valley, was officially opened in 1938 and was used mainly for military purposes during World War II. From 1949, civil and military aviation moved side by side to Caselle, which is now the civil airport of Turin, while the industry, first as FIAT Aviazione, then Aeritalia and now Alenia Aeronautica, has been at Caselle since the early 1950s.

The Alenia Aeronautica Caselle plant, which employs more than 1,700 people (1,000 of whom are working on Eurofighter Typhoon production), is dedicated to the assembly of rear fuselages and port wings, and to the final assembly and flight testing of Eurofighter Typhoon aircraft for the Italian Air Force. The inaugural flight of the first Italian production Eurofighter Typhoon, Instrumented Production Aircraft 2 (IPA 2), took place in April 2002.

Within the Eurofighter Typhoon consortium, Alenia Aeronautica has design responsibility for the wings in collaboration with EADS-CASA of Spain, as well as for the rear fuselage with BAE Systems and for role equipment. The company is also responsible for some important onboard systems, such as armament, navigation, utility control and propulsion, and for the delivery of the entire weapon system to, and its integration into, the Italian Air Force.

Wing assembly and final assembly of Eurofighter Typhoon at Caselle

Eurofighter Typhoon final assembly for the Spanish Air Force at Getafe

EADS-CASA – Getafe, Spain

The EADS-CASA facility at Getafe covers approximately 50 hectares and is shared by the Defence and Security, Airbus and Military Transport Aircraft divisions. Located 13km south of Madrid, Getafe was the site of the capital city's first airport, built in 1921. International flights commenced in 1930. Construcciones Aeronauticas SA (CASA), the forerunner of EADS-CASA, was formed in 1923 and opened a workshop at Getafe in 1924.

The EADS-CASA facility, where more than 700 personnel are directly involved in the programme, is responsible for 13 per cent of all Eurofighter Typhoon production, including the manufacture of the starboard wing for all the aircraft on order. The company also carries out the final assembly of the 87 Eurofighter Typhoons for their national customer; this began in 2001, and the first series production Eurofighter Typhoon for the Spanish Air Force took to the skies from Getafe in February 2003.

The Getafe plant is now also responsible for the maintenance, overhaul and combat efficiency improvement of Spanish Air Force and Navy aircraft such as the AV-8B Harrier, F-18 Hornet, F-5 Tiger and F-1 Mirage, the Tamiz and C-101 Aviojet trainers, the C-130 Hercules transport and the P-3 Orion maritime patrol aircraft.

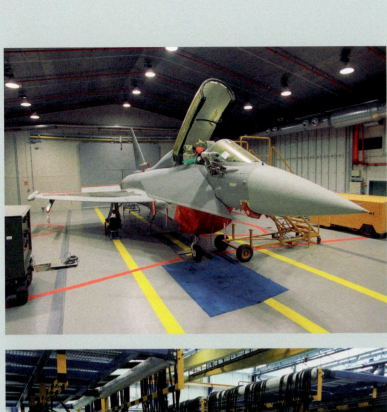

BAE Systems – Warton, United Kingdom

BAE Systems is responsible for 33 per cent of all Eurofighter Typhoon production and employs around 3,000 personnel on the programme across two main sites in the north-west of England, Warton and Samlesbury.

At Warton, BAE Systems carries out the final assembly and flight testing of all Royal Air Force-bound Eurofighter Typhoon aircraft. The airfield, located 11km west of Preston, in Lancashire, functioned as an air depot for the US Army Air Forces during World War II, processing thousands of aircraft on their way to active service in the European theatre. Today, it is a major assembly and testing facility of BAE Systems.

The second main site is at Samlesbury, also in Lancashire, where employees are involved in the manufacture of various Eurofighter Typhoon components, including centre fuselage frames, wing-to-fuselage brackets and foreplanes, and in the assembly of fully equipped front fuselages, including windscreens and canopies, for all aircraft built by the consortium. Additionally, BAE Systems is responsible for the assembly of Stage 1 rear fuselages, which are then shipped to Alenia Aeronautica's plant in Italy for completion.

The origins of this facility date back to 1922, when it was the site of a proposed municipal airfield to serve the nearby towns of Blackburn and Preston. However, construction did not commence until April 1939. Its development was accelerated because of the outbreak of World War II, when the Air Ministry instructed the English Electric Company to proceed with the construction of bomber aircraft.

Eurofighter Typhoon production and final assembly at Warton

Development Aircraft

The EAP (Experimental Aircraft Programme) technology demonstrator in formation with Hawk, Harrier and Tornado aircraft

EAP was produced to test and examine the design features for the potential Eurofighter Typhoon

EAP made its first flight on 8 August 1986

Eurofighter Typhoon is highly manoeuvrable, as demonstrated here by the smoke winders

Eurofighter Typhoon DA 1 and DA 5 in formation over Manching

The first development Eurofighter Typhoon to fly, DA 1 (Development Aircraft 1) initially lacked radar and was fitted with RB199 engines, although the latter were subsequently replaced by EJ200s

DA 1 in the colours of EADS-CASA and the "Bavarian Air Force"

DA 5 taking off with full reheat

DA 1 landing with brake chute

After being fitted with the EJ200, Eurofighter Typhoon DA 1 carried out air-to-ground weapons tests and air-to-air refuelling trials

EADS-built Eurofighter Typhoon taking off at the Berlin Air Show

DA 4, the first example to be fitted with Captor radar and a complete avionics suite

DA 2 painted in high-gloss black to hide pressure transducers

DA 4 after engine shut down

DA 2 ready for flight

DA 4 landing with brake chute

DA 2 pre-flight checks

DA 1, DA 2 and DA 4 in formation en route to the Farnborough Air Show

DA4 air-to-air refuelling from an RAF VC10

DA 2 air-to-air refuelling from a RAF Lockheed Tri-Star tanker. The monitor screen enables the drogue operator to guide the receiving aircraft.

DA 7 with heavy payload

DA 7 on a test flight over Sardinia

DA 3 in formation with an Austrian Saab Draken

DA 3 firing its first Sidewinder missile

Test pilot M. Venanzetti

Chief test pilot M. Cheli

DA 7 over the Sardinian coastline

DA 7 carried out extensive trials with assorted weapons, including Meteor, Storm Shadow, ASRAAM, AMRAAM and Sidewinder

Eurofighter Typhoon is able to carry a wide range of weapons

DA 6 was used for envelope expansion, climatic and Environmental Control System (ECS) testing and Multifunction Information Distribution System (MIDS) and helmet testing, as well as for inflight refuelling

DA 6 on a training sortie

Ferrari and McLaren Formula 1 racing cars with
Instrumented Production Aircraft (IPA)
Eurofighter Typhoons

Development Aircraft and Instrumented Production Aircraft at Warton

A highly capable aircraft: IPA 1 with 1,000lb Paveway II, Sidewinders and drop tank

Test pilot Mark Bowman trials the new helmet

IPA 1 taking off with full reheat

IPA1 taking off with reheat from Warton

IPA 2 flying over the Italian Alps

Sunrise on IPA 2

IPA 2 on a test flight over Sardinia

IPA 2 during a training sortie

IPA 3 with IRIS-T, AMRAAM, Paveway II bombs and drop tanks

IPA 3 over a snow-covered Bavarian landscape

IPA 5 starts its maiden flight from Warton

▲▲ IPA 5 on its first flight at Warton

▲ Alfonso de Castro in the cockpit of IPA 4

Eurofighter Typhoon IPA 4 climbs away from Getafe for a test flight

IPA 4 captured in the early light of day while undergoing environmental trials in Sweden

IPA 4 over the city of Seville

IPA 4 on a test flight over the Spanish coastline

Eurofighter in the German Air Force

On 13 January 2004, the then German Defence Minister, Peter Struck, announced major changes to the German Armed Forces. An important part of this announcement was the plan to cut the fighter fleet from 426 aircraft to 265 by 2015. When the current order for 180 Eurofighters, as the aircraft are known in-service, is fulfilled, 140 will be with operational wings and the German Air Force's Tornado fleet will be reduced to 85.

The first two Luftwaffe (German Air Force) units to be equipped with the Eurofighter are Jagdgeschwader (Fighter Wings) 73 and 74. Fighter Wing 73 "Steinhoff" is based in north-eastern Germany at the town of Laage near Rostock. Its roles include general air defence as well as pilot training for the conversion to Eurofighter. 732 Squadron, part of Fighter Wing 73, is the German Air Force's Operational Conversion Unit.

The first three two-seat Eurofighters arrived at Laage Air Base on 26 April 2004 and a further four had followed by 30 April, the official date of the first flight of a German Air Force Eurofighter.

The last of the wing's former East German Air Force MiG-29s were withdrawn from service on 4 August 2004 and its first single-seat Eurofighter was delivered on 8 May 2005. The wing currently operates ten single-seat and eight two-seat Eurofighters and, as deliveries continue, will eventually swell to its full complement of 36 aircraft.

Fighter Wing 74 is the second German Air Force wing to have received Eurofighter. It is based in southern Germany at Neuburg a. d. Donau, and the first four weapon systems were delivered on 25 July 2006. The wing will take over Quick Reaction Alert (QRA) duties from the F-4F Phantom in January 2008 and will then be on continuous ground readiness. Fighter Wing 74 is also part of the Crisis Reaction Force of the German Air Force.

The German Air Force plans to have replaced all of its F-4F Phantom and parts of its Tornado fleets with the Eurofighter by the year 2013, introducing it to Fighter Wing 31 at Nörvenich, Fighter Wing 71 at Wittmund and, finally, Fighter Wing 33 at Büchel.

The sun sets on Fighter Wing 73 "Steinhoff" at Laage

Pilot in the "office" of an OCU Eurofighter

A pilot assigned to 732 Squadron at Laage dons his life-preserver

A Eurofighter of 732 Squadron, the Operational Conversion Unit, taxies back to "six pack" at dusk

A Fighter Wing 73 Eurofighter in its Laage "six pack" hangar

Three Eurofighters of Fighter Wing 74 in formation

Wingtip ECM pod and IRIS-T missile

Taking off from Laage's runway

Taxiing Eurofighter at Laage

Loading IRIS-T

A Fighter Wing 73 Eurofighter pilot airborne with a two-seater acting as his wingman

A pilot strapping into his Eurofighter

Ground crew checking Eurofighter after flight

Four Eurofighters of 732 Squadron at "Last Chance" before take-off at Laage

A pair of Fighter Wing 74 Eurofighters taxi to the threshold

Ground crew checking IRIS-T after flight

A pair of Eurofighters take off from Laage Air Base in north-eastern Germany

A Laage-based single-seater Eurofighter of Fighter Wing 73 carrying underwing IRIS-T missiles and drop tank

The low sun highlights a Eurofighter at Laage

A hangared Eurofighter at Neuburg

A full afterburner departure from Neuburg Air Base

▲▲ Eurofighters waiting at "Last Chance" for take-off clearance

▲ Crews walking to their Eurofighter aircraft

Eurofighter taxies from shelter at Neuburg

Pilot walks from his aircraft after his training flight

A 732 Squadron Eurofighter with brake chute on landing at Laage

Ground crew prepares for engine start

Pilot climbing aboard his Eurofighter

The EJ200 turbofan is checked by ground crew

Fighter Wing 73 Eurofighter takes off from Laage

Eurofighter pulling high G in afterburner turn

Taking off with afterburners from Laage

A Eurofighter of Fighter Wing 74 taxies from its shelter at Neuburg

Fighter Wing 74 Eurofighter leaves its shelter at Neuburg

A Fighter Wing 73 two-seater Eurofighter breaking upwards on a training sortie

Close up of Eurofighter over Laage

Pilot returns after sortie

Eurofighter cockpit checks

Fighter Wing 73 Eurofighter fins

Ground checks

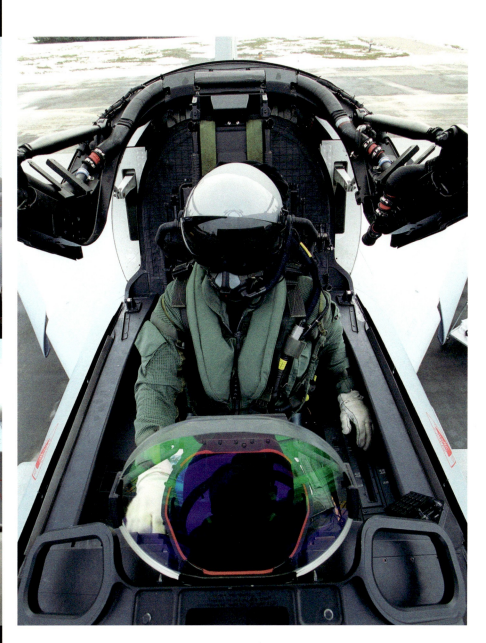

Pilot in twin-seater Eurofighter

Airborne at Laage on 40,000 lb of thrust

▼ Eurofighter pilot on top of the world

▼▼ A pair of Fighter Wing 74 aircraft at Neuburg

Full-afterburner take-off from Laage

Fighter Wing 73 Eurofighter flight line at Laage

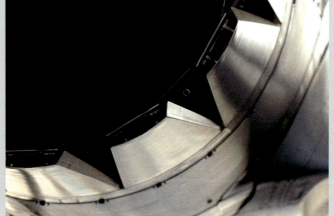
The hot end of an EJ200

A Fighter Wing 73 single-seater with drop tank and IRIS-T missiles

A bird´s eye view of a Fighter Wing 74 Eurofighter as its pilot prepares to depart for a training mission from Neuburg Air Base

A vertical climb by a Fighter Wing 73 Eurofighter high above clouds over northern Germany

A four-ship formation of Fighter Wing 73 Eurofighters

A pair of Fighter Wing 73 two-seater Eurofighters on finals at Laage

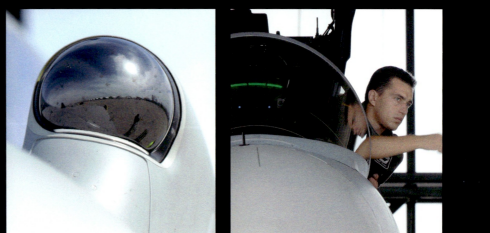

IRST - Infra Red Search and Track

Ground crew working in cockpit

Fighter Wing 73 two-seater flies over the north German countryside

A Fighter Wing 73 two-seater Eurofighter carrying two Air Combat Manoeuvring Instrumentation (ACMI) pods on its underwing pylons

Ground crew checking Eurofighter after flight

Pilot waiting to taxi

Eurofighter returns from sortie at Neuburg

▲▲ Taxiing past the control tower at Laage

▲ Ground crew talking to pilot

Eurofighter returns to "six pack" after sortie at Laage

Eurofighter taxies with landing lights to "six pack" at dusk

Eurofighter in the Italian Air Force

Italy, as one of the partner nations in the Eurofighter programme, will take delivery of 121 aircraft, 15 of which will be the two-seat variant. The last of 27 Tranche 1 Eurofighter Typhoons, designated F2000 in service, will be delivered to the Italian Air Force (Aeronautica Militare Italiana) by the end of 2007, while 52 Tranche 2 aircraft will be delivered between 2008 and 2013.

The first two Italian Air Force units to be equipped with the F2000 are IX° and XX° Gruppos (Squadrons), which are part of the 4° Stormo (Wing) at Grosseto Air Base, 185km north-east of Rome. The first two-seat F2000 arrived at Grosseto on 18 March 2004 and the type has gone on to replace the wing's veteran F-104 Starfighters. XX° Gruppo is the F2000 Operational Conversion Unit (OCU), while IX° Gruppo is the Italian Air Force's first front-line unit.

On 16 December 2005 the aircraft took over Quick Reaction Alert (QRA) duties and carried out their first high-profile operations by protecting the skies over Torino (Turin) during the 2006 Winter Olympics. Initial Operational Capability (IOC) was declared on 1 January 2007 and Full Operational Capability (FOC) is expected to be achieved by the end of 2008. The F2000 will take over full responsibility for the air defence of Italy from the F-16A-ADFs that have been leased from 5° Gruppo, based at Cervia near Rimini, when the latter are withdrawn from service in 2010.

The latest unit to be equipped with the F2000 is 12° Gruppo, part of 36° Stormo based at Gioia del Colle in southern Italy. Two of the wing's other squadrons will receive F2000s following the signature of the Tranche 3 contract for a further 46 aircraft.

A 4° Stormo pilot carries out pre-flight checks on his F2000 at Grosseto

A pair of Italian Air Force 4° Stormo F2000s flying near Torino on QRA duty

An F2000 taxies outside a shelter at Grosseto

Two single-seater F2000 aircraft assigned to the 4° Stormo at Grosseto Air Base

Two F2000s of 4° Stormo, the first front-line Italian Air Force squadron to be equipped with the aircraft

The F2000's Infra Red Search and Track (IRST) system

Pilots settle into the two-seater F2000 cockpit

A Block 5 F2000 taking off at Grosseto

A pilot pre-flight checks his missiles

Three of 4° Stormo's single-seater F2000s in close echelon formation

Italian Air Force F2000 with the sun low over the sea

An Italian Air Force F2000 over the mountains near Torino

F2000 fins on the flight line

Ground crew guide an F2000 to its parking position

Three F2000s in "vic" formation over the Tuscan countryside

single-seater F2000 lines up for take-off from Grosseto

A single-seater F2000 turns on the power

+++ above the rest +++ Eurofighter Typhoon +++

Block 5 F2000s in formation

An F2000 taxies to the runway

A Grosseto crew chief speaks to the ATC

F2000s of IX° Gruppo took over Quick Reaction Alert (QRA) duties in Italian airspace in December 2005

A pair of IX° Gruppo F2000 aircraft on patrol over a hazy landscape

A IX° Gruppo pilot in the "office" of his F2000 single-seater, firmly strapped into his Type 16A ejection seat

An F2000 of 12° Gruppo – part of 36° Stormo based at Gioia del Colle

Pilots at a briefing sortie

Pilots prepare for flight

A pair of IX° F2000s breaking for action

The pilot's view from a 4° Stormo F2000

A Gosseto ground crewman speaking to his pilot during pre-flight checks

Sunshine on a 4° Stormo F2000

The XX° Gruppo flight line of single-seater F2000s at Grosseto

F2000 silhouetted by the setting sun

F2000 two-seaters return from a training mission at dusk

Eurofighter in the Spanish Air Force

The Spanish Air Force (Ejército del Aire Español) is currently replacing the older combat aircraft on its inventory with newer types, including the Eurofighter (as it is known in the Air Force ranks). Spain, as one of the partner nations, has accepted into service all nineteen Eurofighter aircraft from the Tranche 1 buy and issued them to Ala (Wing) 11, based at Morón de la Frontera near Seville. They consist of eleven single-seaters, designated C.16, and eight two-seaters, designated CE.16.

The Morón wing comprises three Escuadrones (squadrons), 111 and 112, which will be on combat duty, and 113, which will be used for operational training. The first Spanish two-seat Eurofighter, ST001, was delivered on 4 September 2003, and, in the build-up of operations since then, Ala 11 has achieved Initial Operational Capability (IOC). Full Operational Capability (FOC) will be achieved in spring 2008.

The 35 aircraft of the second tranche are due to be delivered in 2008. These Eurofighters will be issued to Ala 11 and, later, to Ala 14 based at Los Llanos near Albacete, replacing Mirage F1s.

The Spanish Air Force has a total requirement for 87 Eurofighter weapon systems.

A Spanish student pilot and his instructor leave their Eurofighter in its shelter at Morón Air Base

Eurofighter pilot's head gear

The Spanish Air Force's Wing 11 operates eight two-seater Eurofighters at Morón Air Base

An instructor pilot briefs a student pilot prior to a Eurofighter training mission

A Wing 11 pilot pre-flight checks his two-seater aircraft at Morón

Worn with pride

Spanish Eurofighter fins

One of eleven single-seater
Eurofighters based at Morón

The powerful Spanish sun glints off
a Eurofighter's head-up display

A two-seater Wing 11 Eurofighter in a full afterburner climb out from Morón Air Base

Standing out against the clear blue sky of southern Spain, this Wing 11 Eurofighter shows its foreplanes and leading edge slats

Wing 11 pilots head for a debrief at Morón Air Base after a training mission in their Eurofighter

A Wing 11 Eurofighter lifts off over CASA Aviojets belonging to the Spanish Air Force aerobatic team

A Spanish single-seater Eurofighter - the star of the air show

Pilots climb aboard a two-seater Eurofighter

A Eurofighter taxies from its shelter at Morón

The concentrated power from Spanish Air Force Eurofighters

The Spanish Air Force's two-seater Eurofighters, designated CE.16, are operated by Wing 11's 113 Squadron at Morón Air Base

Wing 11 Eurofighters overfly Morón Air Base returning from a training mission

Pilot climbing into Eurofighter cockpit

▲▲ Head-on view of a Spanish Air Force Eurofighter

▲ Cleared for engine start

A Wing 11 Eurofighter belonging to 113 Squadron, the Operational Conversion Unit (OCU) based at Morón, flies over the rugged terrain of southern Spain

With his instructor in the rear seat, a student pilot flies in close formation as part of his training mission in a 113 Squadron CE.16 aircraft

Gear down – power on

A Wing 11 two-seater Eurofighter makes a high-G turn above the misty hills

Typhoon in the Royal Air Force

Named Typhoon in the United Kingdom, Eurofighter entered service with the Royal Air Force late in 2003. The first squadron, No 17 (Reserve), which is also the Operational Evaluation Unit, has been busy since then assessing weapon systems performance, trialling new equipment and developing tactics. No 29 (Reserve) Squadron, the Operational Conversion Unit, has had the responsibility for the initial training of the air and ground crews for the operational squadrons.

RAF Coningsby was chosen to become the home of the Typhoon F.2, and two operational squadrons, Nos 3 (Fighter) and XI Squadrons, together with No 17 and No 29, are based at the famous Lincolnshire airfield. No XI, currently in the process of building up to full strength, is the lead unit for those Typhoons tasked with ground attack and is expected to be fully operational in both the air-to-air and the air-to-ground roles during the course of 2008. The first two multi-role Block 5 Typhoons were delivered to No XI Squadron at RAF Coningsby on 6 August 2007.

In addition to its range of air-to-air weapons, the multi-role Typhoon's initial air-to-ground weapons package will include the free-fall 1,000lb bomb, the Paveway II (PW II) precision-guided bomb, the Enhanced PW II with improved GPS guidance, and the Mauser 27mm cannon. The first drop trial of an Enhanced PW II using a laser designator took place on 7 August 2007. All Royal Air Force Typhoons will eventually be capable of operating in both the air-to-air and the ground attack roles.

No 3 (Fighter) Squadron, which received its first Typhoon in March 2006, is the lead squadron for the Royal Air Force's air defence task. This squadron became the first front-line Typhoon unit to fire the Advanced Short-Range Air-to-Air Missile (ASRAAM), an event that took place at the Aberporth weapons range over Cardigan Bay, Wales, during the week beginning 26 February 2007 and proved the Typhoon's outstanding capability in this role.

Typhoon F.2s are equipped with a highly advanced avionics suite, the heart of which is the ECR 90 multi-mode radar that incorporates Non Co-operative Target Recognition. Typhoons are also equipped with an Infra Red Search and Track System, a Multifunction Information and Distribution System data link and a fully automatic Defensive Aids Sub-System. The aircraft has a wide-angle head-up display that will shortly be complemented by a helmet-mounted display; both of these project flight reference data, weapon aiming and cueing, and forward-looking infra-red imagery. The cockpit also contains three multifunction, full colour, head-down displays and is compatible with the use of night-vision goggles.

Typhoon F.2s took up their operational duties on 29 June 2007 when they assumed responsibility for part of the Quick Reaction Alert (QRA) element of United Kingdom's air defence in the south of the country at RAF Coningsby. QRA procedures entail aircraft being held at continuous ground readiness so that they can take off within minutes, at any time, to protect the skies over the nation. During the course of the next nine months, Typhoons will progressively take over the entire southern QRA as they replace the Tornado F.3s that have performed this duty for many years. The first interception of a Russian Air Force aircraft by an RAF Typhoon from Coningsby on QRA duty took place on 17 August 2007 when a Tu-95 "Bear-H" bomber approaching United Kingdom airspace was shadowed by an F.2 from No XI Squadron over the North Atlantic Ocean.

Under current Ministry of Defence planning, the northern QRA at RAF Leuchars, near St Andrews, Scotland, will continue to fly Tornado F.3s until the squadrons based there re-equip with Typhoons later in the decade. The Royal Air Force is expected to deploy the multi-role Typhoon on overseas operations in 2008, possibly to Afghanistan.

A No 29 (R) Squadron single-seater Typhoon F.2 climbing above the clouds

Ground crew check Typhoon "Alpha Bravo"

An RAF ground crewman waits for two Typhoons to taxi to the apron at RAF Coningsby

A line-up of No 17 (R) Squadron Typhoon T.1 noses

A formation of three two-seater Typhoon T.1s of Nos 17 (R) and 29 (R) Squadrons

No 17 (R) Squadron Typhoon markings

A pair of Typhoons parked under the floodlights at Warton

Ground crewmen attend to a Typhoon

The drooped foreplanes of a Typhoon at rest

A pair of No 3 (F) Squadron Harrier GR.7s break away from a Typhoon F.2, their successor aircraft

A No 29 (R) Squadron Typhoon T.1 seen from the cockpit of its wingman

The view forward from the rear seat of a Typhoon of No 29 (R) Squadron

Formation of five No 17 (R) Squadron Typhoon F.1s as one breaks away

A Typhoon pilot prepares for his mission at RAF Coningsby

A pilot carrying out pre-flight checks

A ground crewman checks a wing-mounted ASRAAM

Ground crew check a No 17(R) Squadron Typhoon cockpit

Typhoon air-to-air refuelling from an RAF VC10 tanker aircraft

Typhoon breaking into the airfield circuit

A No 3 (F) Squadron Typhoon pilot straps himself into his cockpit

A Typhoon pilot climbs out of his aircraft after the last mission of the day

Sunset taxi at RAF Coningsby

A No 17 (R) Squadron Typhoon carrying ASRAAM and Meteor Beyond Visual Range air-to-air missiles

A No 29 (R) Squadron Typhoon in afterburner take-off

Ground crew checking ejection seat straps

An open canopy highlighted against a No 17 (R) Squadron Typhoon

A Typhoon pilot carries out weapon checks on an ASRAAM

A No 29 (R) Squadron pilot talks to ground crew before starting engines

Twin EJ200 bypass turbofans with afterburners power a Typhoon skyward

No 3 (F) Squadron Typhoon F.2s took over the Quick Reaction Alert (QRA) responsibilities for the air defence of the United Kingdom in June 2007

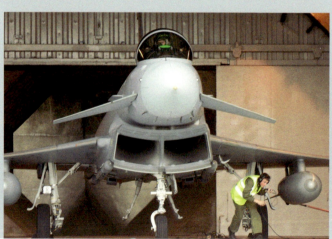

Final check by ground crew before a No XI Squadron Typhoon taxies

▲▲▲ A No XI Squadron Typhoon pilot strapped into his Type 16A ejection seat

▲▲ No 3 (F) Squadron crewmen sprint towards a Hardened Aircraft Shelter (HAS) on a QRA scramble

▲ Ready for action: a QRA Typhoon F.2 of No 3 (F) Squadron

A No 3 (F) Squadron Typhoon F.2 breaks away during a QRA sortie

A No 17 (R) Squadron Typhoon F.2 in formation with an Indian Air Force Su-30 MKI during Exercise "Indra Dhanush"

EJ200 engine afterburners light up the darkening sky

RAF Typhoons in "Diamond Nine" formation

Tailfins of Nos 29 (R) and 17 (R) Squadron Typhoons

▲▲ The last mission of the day

▲ A Typhoon of No XI Squadron in silhouette

Nos 3 (F) and 17 (R) Squadron Typhoons in echelon starboard formation

A "vic" formation of three No 3 (F) Squadron Typhoons

A No XI Squadron pilot walks to his aircraft

The high cockpit of the Typhoon gives the pilot an excellent view

A No XI Squadron pilot waits for taxi clearance

Nine F.2 Typhoons in "Typhoon" formation

No 17 (R) Squadron Typhoon F.2s leave contrails as they formate on their leader

A No 29 (R) Squadron Typhoon F.2 in a vertical climb

Ground crew loading AMRAAMs on to a No 17 (R) Squadron Typhoon

A No 3 (F) Squadron Typhoon fires an Advanced Short-Range Air-to-Air Missile (ASRAAM)

A No 29 (R) Squadron Typhoon F.2 on a training sortie

The Red Arrows fly a "Heart" manoeuvre over a No 3 (F) Squadron Typhoon at RAF Fairford

▼ Silhouette of a No 29 (R) Squadron Typhoon at RAF Fairford

▼▼ Ready for night flight

A No 29 (R) Squadron Typhoon

A No XI Squadron Typhoon pilot walks to his aircraft

A No 17 (R) Squadron Typhoon F.2 over the Lincolnshire countryside, fully armed with ASRAAM and AMRAAM, Paveway II bombs and a drop tank

A No 3 (F) Squadron pilot going through pre-flight checks prior to start-up

A No 29 (R) single-seater Typhoon F.2 with afterburners glowing

Instructor and student pilot after completing a training mission in an Operational Evaluation Unit (OEU) Typhoon T.1

The upper wing surface of the Typhoon T.1

Brake chute safety pin

Close formation with an RAF Tornado F.3, a No 17 (R) Squadron Typhoon F.2 and an Indian Air Force Su-30 MKI

Typhoon "Diamond Nine" formation

A No 29 (R) Squadron Typhoon in the vertical

Ground-crew check of a No XI Squadron Typhoon's wingtip ESM/ECM pod

A No 29 (R) Squadron Typhoon F.2

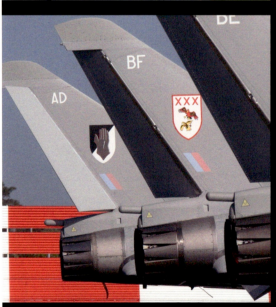

RAF Coningsby OCU and OEU fins

A No XI Squadron Typhoon – one of the latest Block 5 aircraft

A No 17 (R) Squadron Typhoon leaves its Hardened Aircraft Shelter

A pair of No 29 (R) Squadron two-seater Typhoon T.1s

Over the North Sea: a No 17 (R) Squadron Typhoon on a training mission

+++ above the rest +++ Eurofighter Typhoon +++

A mixed formation of No 3 (F) and No 29 (R) Squadron Typhoons

▼ The first Block 5 aircraft leaves Warton

▼▼ The return to base

▲▲ A No 29 (R) Squadron Typhoon

▲ No 29 (R) Squadron ground crew use a Typhoon to shelter from the rain

A "Box Four" formation of Typhoons leads the Red Arrows over the Lancashire coastline

A No 3 (F) Squadron Typhoon practises "touch and go's" at RAF Coningsby

A No XI Squadron Typhoon with brake chute deployed

A No 3 (F) Squadron Typhoon F.2 at low level among the Welsh valleys

Eurofighter in the Austrian Air Force

Following protracted negotiations between the new Government, elected in October 2006, and the industrial consortium, Austria became the first export customer for the Eurofighter Typhoon when an order for fifteen single-seat aircraft was confirmed. In service, the weapon system is referred to simply as "Eurofighter", and the aircraft will replace ten F-5Es leased from Switzerland from 2008.

The unit operating Eurofighter will be the Austrian Air Force's Überwachungsgeschwader (Air Surveillance Wing) based at Zeltweg/Steiermark. The wing belongs to the Air Surveillance Division, which is stationed at Salzburg and is responsible for the air defence of Austria. The unit received its first Eurofighter aircraft in July 2007 and its second in September, with deliveries set to be completed in 2008.

The conversion to type for pilots and ground technicians has been conducted in co-operation with the German Air Force. The first six Austrian pilots attended their conversion training during the second half of 2006 and were assigned for practical flight training with Fighter Wing 73 of the German Air Force at Rostock-Laage from 15 January 2007. Conversion to type training, which includes 30 simulator flying hours and 27 real flying hours, runs over five months.

The first Austrian maintenance engineers have completed their Eurofighter training. A further 20 began their conversion in September 2006 with industry, while, at the same time, 70 more personnel started working with the German Air Force at the latter's Technical School in Kaufbeuren.

One of the Austrian Air Force's single-seater Eurofighters is towed from its hangar at Zeltweg

▲▲ An Austrian ground crewman waves in the first Eurofighter

▲ Opening one of the aircraft's stowage bays

The first Austrian Air Force Eurofighter landing at Zeltweg Air Base in July 2007

Eurofighter being welcomed on its first arrival at Zeltweg

The Austrian Air Force Eurofighter markings

Eurofighter ready to taxi

An Austrian Air Force F-5E formates with Eurofighter

Pilot walking to his Eurofighter

Pilot signing for his Eurofighter before flight

▲▲ Austrian Air Force F-5Es formate with Eurofighter

▲ An Austrian Air Force pilot pre-flight checks his Eurofighter

The first Austrian Air Force Eurofighter, escorted by a pair of F-5Es

Ready for the first flight

Lining up for first departure

Eurofighter towed from its hangar before its first flight from Zeltweg

An Austrian Air Force Eurofighter lifts off the runway at Zeltweg for the first time

Pilot checks the missile pylon of his Eurofighter at Zeltweg

Taxiing to the runway for the first flight of an Austrian Air Force Eurofighter from Zeltweg

The Austrian Air Force single-seater Eurofighter will be operated by its Air Surveillance Wing based at Zeltweg Air Base

Air brake extended, Eurofighter lands back at Zeltweg after its first sortie

The Austrian ground crew greet "Seven-Lima" after its first flight from Zeltweg

The pilot climbs aboard Eurofighter

Eurofighter in German Air Force markings prior to its delivery to the Austrian Air Force

An enthusiastic "thumbs up" from the delighted Austrian Air Force Eurofighter pilot

Landing back at Zeltweg after its first sortie

In pristine condition, the first Austrian Air Force Eurofighter takes to the skies

An Austrian Air Force Eurofighter offering a perfect illustration of the aircraft's plan view

The sun hits the upper surfaces of a German-marked Austrian Air Force Eurofighter as it heads for a cloudbank

Pilot opens the canopy after a training sortie

Ground crew checking the Eurofighter on-board systems and the nose wheel

External power added

A ground crewman cleans the canopy

Inserting the "Remove Before Flight" safety pins

Blanks being fitted at the end of a day's flying

Back to the hangar

An Austrian Air Force Eurofighter wrapped and tagged at Zeltweg Air Base

Acknowledgements

For this volume, the author would like to thank W Hoeveler, VP Communications, and his team at Eurofighter GmbH in Hallbergmoos, and the test pilots, engineers and PR/communications teams at Alenia Aeronautica S.p.A in Rome, Caselle and Turin Plant, at EADS Deutschland in Munich, at EADS CASA in Getafe, at Eurojet GmbH in Munich and at BAE Systems at Warton.

He is indebted also to the pilots, ground crews and PR/communications teams of the German Air Force's JG 73 at Rostock/Laage and JG 74 at Neuburg, of the Italian Air Force's 9 and 20 Gruppos at Grosseto, of the Spanish Air Force's Ala 11 at Morón, of the Austrian Air Force at Zeltweg and of the Royal Air Force's Nos 3 (F), XI , 17 (R) and 29 (R) Squadrons at RAF Coningsby. All have welcomed him to their bases to record his images.

A special "thank you" is offered to No 100 Squadron at RAF Leeming, who have provided photo chase aircraft to enable the author to capture some of the images.

Photo Credits

The majority of the photographs in **Above the Rest** were taken by the author, but he would like to thank those organisations and individual photographers listed below for permitting the use of their superb material, all of which has contributed greatly towards making this volume on Eurofighter Typhoon the most comprehensive to date.

Alenia Aeronautica: 12, 13, 42, 43, 48, 114

BAE Systems: 10, 27
Ray Troll: 5, 7, 11, 12, 24, 25, 36, 49, 50, 52, 60, 172
Chris Ryding: 34, 39, 50, 145, 173

EADS-CASA: 30, 32, 46, 47, 62, 121,129, 130, 131,132, 135

EADS Deutschland: 12, 18, 29, 30, 185, 188

Eurojet GmbH: 17

Austrian Air Force
Markus Zinner: 176, 177, 178, 179, 180, 181, 182, 184, 186, 187, 188, 189, 190

German Air Force
Peter Steiniger: 69, 72, 77, 83, 84, 89, 91, 94, 95, 96
Ralf Brandis: 88

Italian Air Force
Troupe Azzurra: 103, 104, 108, 112, 113, 115
Laboratorio Fotografico 4° Stormo: 103, 106, 109, 110, 111, 116, 118

Freelance Photographers
Tom Hill: 175
Patrick Hoeveler: 68, 70
Jamie Hunter: 142, 156, 162
Dr R Niccoli: 107, 119

Edited by David Oliver and François Prins